Food 118

肥美的鸡肉

Big, Fat, Tasty Chicken

Gunter Pauli

冈特·鲍利 著

凯瑟琳娜·巴赫 绘

颜莹莹 译

特别感谢以下热心人士对译稿润色工作的支持：

王必斗　王明远　王云斋　徐小怙　梅益凤　田荣义
乔　旭　张跃跃　王　征　厉　云　戴　虹　王　逊
李　璐　张兆旭　叶大伟　于　辉　李　雪　刘彦鑫
刘晋邑　乌　佳　潘　旭　白永喆　朱　廷　刘庭秀
朱　溪　魏辅文　唐亚飞　张海鹏　刘　在　张敬尧
邱俊松　程　超　孙鑫晶　朱　青　赵　锋　胡　玮
丁　蓓　张朝鑫　史　苗　陈来秀　冯　朴　何　明
郭昌奉　王　强　杨永玉　余　刚　姚志彬　兰　兵
廖　莹　张先斌

目录

Contents

一只公鸡站在一棵栗子树的树枝上，俯视着他的母鸡群。他为她们感到骄傲，也为自己在禽类中的古老血统而自豪。他发现一只黄蜂正在筑巢，便开始向她抱怨起鸡饲料来。

“这儿的食物真糟糕。”公鸡说。

“你说什么？”黄蜂问，“你的主人不让你吃饱吗？”

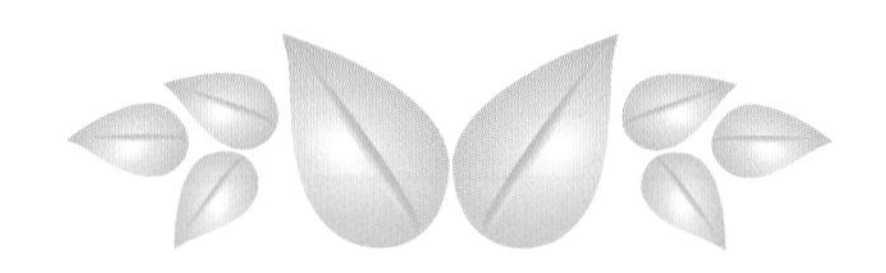

A rooster oversees his flock of hens from the branch of a chestnut tree. He is proud of them, and of his ancient lineage in the bird world. He notices a wasp building its nest and starts a conversation, complaining about the food the chickens are fed.

“The food here sucks,” comments the rooster.

“What do you mean?” the wasp asks. “Does your farmer not give you enough to eat?”

一只公鸡俯视着他的母鸡群……

A rooster oversees his flock of hens ...

他想要的只是我的鸡肉而已！

All he is after is my meat!

“我是说，这些从很远的地方运来的人工饲料，吃起来太乏味了。他用大量的人工饲料来喂养我，让我尽可能快地增肥，这样他就能赚更多的钱。他想要的只是我的鸡肉而已！”

“那么，农民还能指望从你，一只公鸡身上得到什么呢？鸡蛋？”

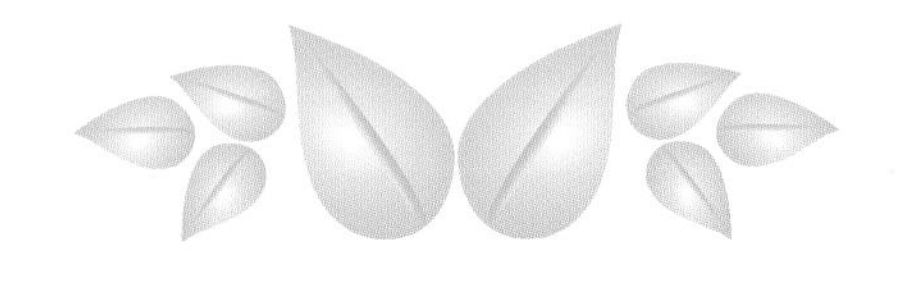

“You know, this processed chicken feed, that is brought here from far away, tastes pretty bland. And he feeds me lots of it – just to get me fat as possible as fast as he can – so he can make more money. All he is after is my meat!”

“Well, what else can the farmer expect from you, a rooster: eggs?”

“不，母鸡才下蛋。鸡农总是将鸡崽分开喂养，一些是用来产肉的，剩下的是用来产蛋的。”

“你是指下蛋的那些母鸡都皮包骨头，所以人们没法吃她们的肉吗？”

“农民是想让那些母鸡尽可能多地下蛋，所以他用另一种饲料喂养她们，好让她们每天都能下一个蛋。”公鸡解释道。

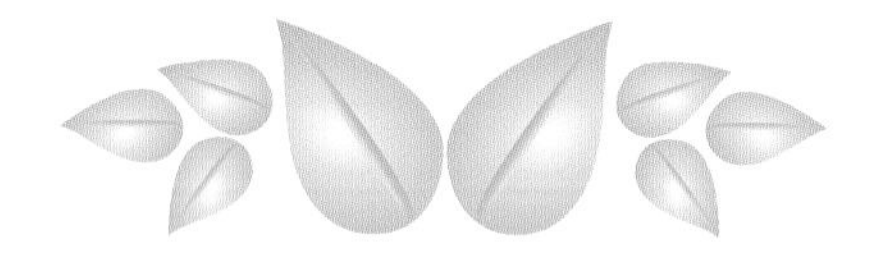

“No, the hens provide that. Chicken farmers always separate the chicks, some for producing meat and the others to lay eggs.”

“Do you mean that the laying hens, with no meat on their bones, are too skinny for people to eat them?”

“The farmer wants those hens to produce as many eggs as possible, so he feeds them a different feed, so they can each pop an egg a day,” Rooster explains.

母鸡尽可能多地下蛋……

Hens to produce as many eggs as possible ...

鸡汤……

Chicken soup ...

“一天下一个蛋！那 定很累。难怪母鸡每次下蛋都会叫个不停。”

“还有更糟糕的：如果她不能一天下一个蛋，她就会被宰杀，一刻也不耽搁。”

“你的意思是，你为农民辛苦工作一辈子，但是一旦你不能每天都下蛋，你就得变成鸡汤？”

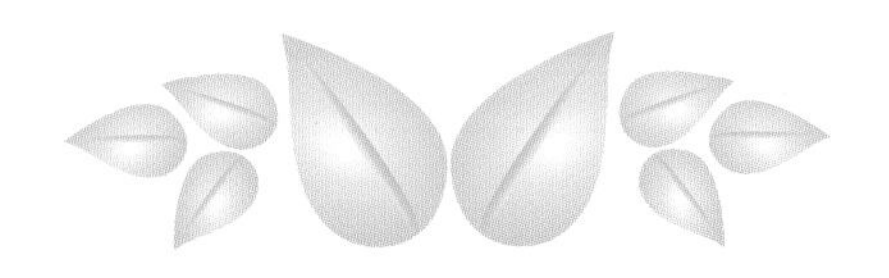

“An egg a day! That must be exhausting. No wonder a hen makes so much noise every time she lays an egg.”

“And worse: if she does not produce an egg a day, she is slaughtered, without delay.”

“You mean to say that when you have worked hard for the farmer all your life, the moment you no longer produce an egg a day, you are turned into chicken soup?”

“是的，大多数母鸡一旦不下蛋了，就会变得一文不值。”

“一文不值吗？”黄蜂问道。

“是的！有些人只知道追求利益，我的朋友。他们使用的强效饲料可以让一只缺乏运动的小鸡在四十天内就变成一只大肥鸡。”

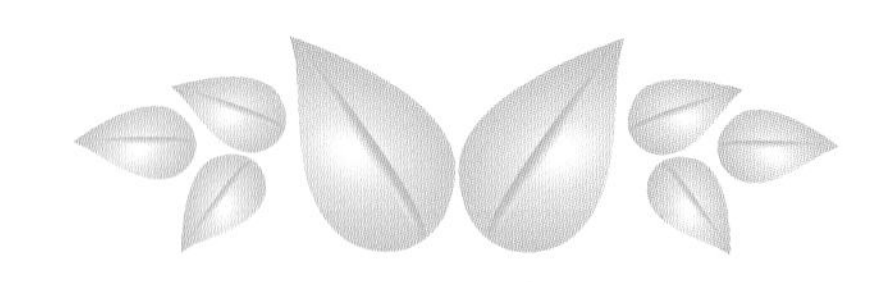

“Well, most of my hens – the minute they stop laying – end up as dog food.”

“Dog food, really?” Wasp asks.

“Yes! For some people it is all about profit, my friend. With the power feed they give them, and the lack of exercise, a tiny chick can be turned into a fat bird within just forty days.”

大多数母鸡变得一文不值

Most end up as dog food

他们被空运过来，然后再用卡车运输……

Flown in by plane, delivered by truck ...

“你在开玩笑！没有人可以长得那么快。”黄蜂说。

“还有更糟糕的，那些小鸡刚到农场时仅仅刚出生一天，他们被空运过来，然后再用卡车运输。这些人工孵化出来的小鸡甚至从来都没有见过他们的妈妈！”

“不！你需要采取行动，公鸡先生。你不能只坐在那里抱怨。”

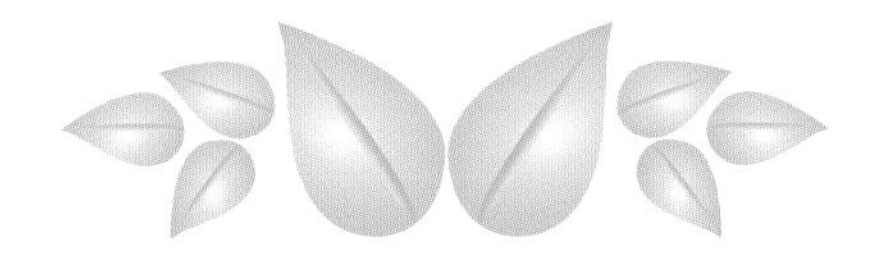

“You’ve got to be joking! No one can grow that fast,” Wasp says.

“What is even worse is that the little chicks that join us on the farm, are only one day old when they arrive, flown in by plane, and then delivered by truck. Those hatchlings have never even met their mothers!”

“No! You need to take action, Mr Rooster. You cannot just sit there and complain.”

“好吧，你身上带刺，所以说来容易。你可以很快行动，但我却得多花点时间。我是有着古老血统的禽类，有着引以为豪的出身和遵循自然的生活方式。我们不想仅仅因为那些银行家跟投机商总是追求快速获利而失去这些。确实需要改变了……”公鸡说。

“那么告诉我，你打算怎么做呢？”黄蜂追问。

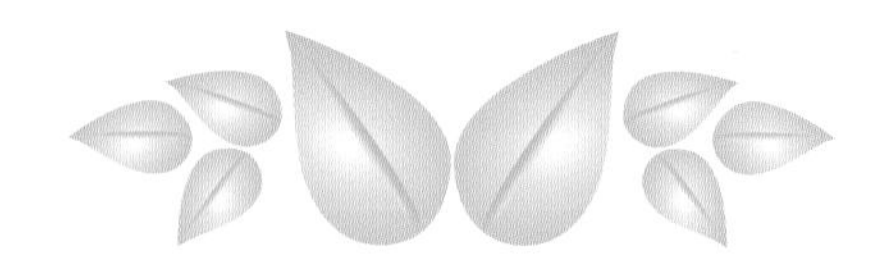

“Well, with your stinging power it is easy for you to say. You can take quick action, for me it takes a bit longer. I am from an ancient lineage of birds and very proud of my chicken heritage, and our natural way of life. We don't want to lose that just because bankers and investors are always after a quick buck. Things have to change...” Rooster says.

“So tell me, what are you going to do about it?” Wasp challenges him.

有着引以为豪的出身……

Very proud of my heritage ...

每天有一定量的海藻

A daily portion of seaweed

“我很清楚要怎么做。首先，我会要求用天然饲料来喂养我们：虫子，种子，还有昆虫，比如蟋蟀。我还会要求每天都有一定量的海藻。”

“我知道海边有许多海藻，但是用海藻做鸡饲料？你可真会开玩笑。”

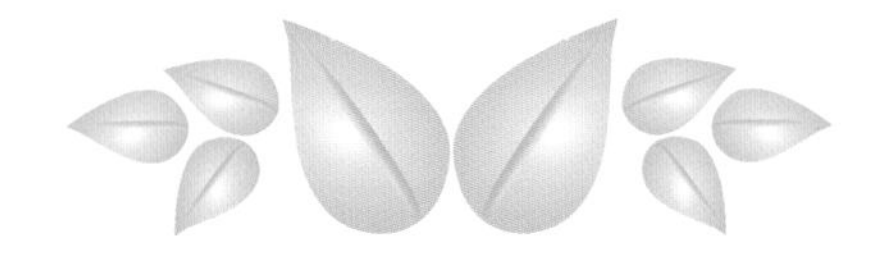

“I know exactly what I am going to do: I will start by requesting that we are fed natural food again: worms, seeds, and insects, like crickets. As well as a daily portion of seaweed.”

“I know there is a lot of seaweed along the coast, but feeding it to chickens? You’ve got to be kidding.”

“绝对没开玩笑。海藻可以使我们的蛋黄呈明亮的橘色，还可以使我们的肉富含碘。把鸡蛋混合入牛奶或羊奶中可以使冰淇淋格外美味。如果我们吃得好，汲取自然的馈赠，那么其他人也会吃得很好。”

“那样的话大家都受益！”黄蜂欢呼道。

……这仅仅是开始！……

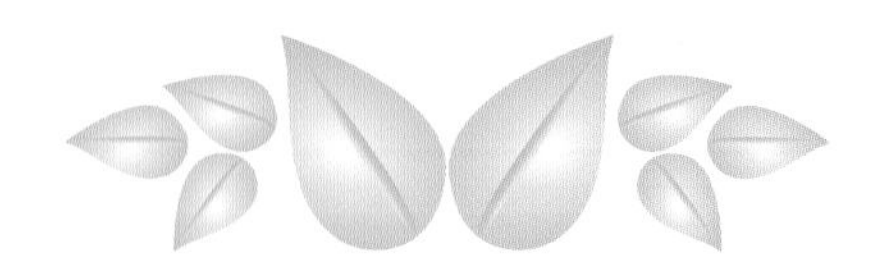

“Absolutely not. Eating seaweed makes our yolks a bright orange and our meat rich in iodine. Adding our eggs to milk from cows or goats makes ice cream extra tasty. If we eat well – from Nature’s bounty – others will eat well too.”

“That way everyone benefits!” Wasp exclaims.

... AND IT HAS ONLY JUST BEGUN!...

……这仅仅是开始！……

... AND IT HAS ONLY JUST BEGUN! ...

Did You Know?

你知道吗？

A normal, healthy chicken, whose ancestors have roamed freely for the past 5 000 years, needs about six months to reach a weight of 2-3 kilograms. A broiler chicken reaches maturity (for slaughtering) after 40 days, by having had a special gene introduced that promotes obesity among birds.

鸡的祖先早在5000年前就开始自由走动了。一只正常、健康的鸡增重2—3公斤需要约6个月。一只肉食鸡通过注入一种特殊基因，只需40天即可长成（用于屠宰），这种基因能促使鸟类变得肥胖。

99% of all chicken meat comes from industrially farmed chickens. As these chickens do not move around at all and have food available 24 hours per day, these birds gain weight fast.

99%的鸡肉来自工业化饲养的鸡。这些鸡不需要走动，一天24小时食物供给不断，所以能够很快增重。

A modern-day farmed chicken has up to ten times more body fat than chickens in Grandmother's days. A serving of chicken has twice the amount of fat of an ice cream.

一只现代化养殖的鸡要比我们祖母那个时代的鸡肥胖 10 倍。一顿鸡肉餐的脂肪含量是一支冰淇淋的 2 倍。

Chicken meat has lost its status as a food that is low in fat and healthy for the heart. Broiler chickens require 2 kg of feed for every kilogram of meat they produce. As soon as a bird reaches its target weight, it must be slaughtered or it will die of heart failure due to excessive growth.

鸡肉已不再是脂肪含量低、有益心脏健康的食品了。肉食鸡每增加 1 公斤重量就需要 2 公斤的饲料。一旦一只鸡达到其目标体重，就必须被宰杀，不然它就会死于过度生长导致的心脏衰竭。

Seaweed is rich in iodine. This is not only healthy, but indispensable for the proper function of the thyroid gland. Eggs from chickens fed seaweed have bright orange yolks, making it look more appealing to the consumer.

海藻富含碘。这不仅有益健康，而且是甲状腺功能正常运作所必需的。鸡吃了含有海藻的饲料，蛋黄会呈明亮的橘黄色，看上去更吸引顾客。

Egg enhance the taste of ice cream. Egg yolk contains a fat that when frozen contains less ice crystals than water. Eggs contain lecithin, an emulsifier that offers body and richness to, for instance, baked goods, sauces and mayonnaise.

鸡蛋可以提升冰淇淋的口感。蛋黄中含一种脂肪，冷冻时的冰晶比水更少。鸡蛋中的卵磷脂可以作为乳化剂使食物的口感更为浓郁和丰富。

Egg improve the stability of an ice cream, reducing its tendency to melt. It also prolongs the shelf life and maintains softness on refreezing. Egg yolk enhances the flavour of vanilla and chocolate.

鸡蛋能够提高冰淇淋的稳定度，使其不会融化得那么快。还可以延长保质期，并且在重新冷冻时保持冰淇淋的软度。蛋黄可以提升香草和巧克力的味道。

Goat's milk has smaller fat globules resulting in a softer curd. This makes goat's milk ice cream tastier and softer on the palate. Goat's milk contains less lactose for those who are lactose intolerant.

羊奶中的脂肪球较小，凝乳更软。这使得羊奶冰淇淋的口感更美味顺滑。对于乳糖不耐的人来说，羊奶中乳糖含量较低。

Which do you prefer: chickens that get fat and meaty fast, or ones that grow naturally?

你会选择快速增肥的鸡还是那些自然生长的鸡？

Do you think of chickens as just providing either eggs or meat? Or do you think that chickens that produce eggs should also be used for their meat when they reach maturity?

你认为可以只用产蛋或者产肉来区分鸡吗？还是说你觉得蛋用鸡长成后也可以肉用？

Do you think that chicken should be used in animal feed?

你认为鸡肉可以用作动物饲料吗？

Do you enjoy only eating ice cream made with egg, or are you ready to try ice cream made without any milk or eggs?

你只喜欢吃用鸡蛋做的冰淇淋，还是愿意尝试不含任何蛋、奶成分的冰淇淋？

Ask your friends and family members if they enjoy eating chicken. Ask if they like chicken broth. Now tell them more about the chicken they enjoy eating at home or in restaurants. Do they have any idea how fast the chicken had grown and how short its life was before if was turned into a "chicken wing"? Inform people about the knowledge you gained from reading this fable and question if taste is more important than the chicken's quality of life. Would your friends prefer eating broiler chickens that are cheap to produce, and tasty only due to special feed and unique genetics? Or do they prefer free-roaming organic chickens that grew up in a natural environment, walking about freely and foraging for seeds, bugs and food scraps from the kitchen? Compare notes with your friends.

问一问你的朋友和家人是否喜欢吃鸡肉，是否喜欢鸡汤。如果他们喜欢在家或在餐厅享用鸡肉，那么跟他们分享一下你对鸡的了解。他们是否知道鸡生长得有多快？在成为“鸡翅”这道菜之前它的生命有多短？让他们了解你从这个故事里学到的知识。问问他们，与鸡的生活质量相比，是否味道更重要？与你的朋友一起对比一下肉食鸡与天然生长的鸡：肉食鸡价格低廉，味道可口，使用特殊饲料喂养，有着特殊品种；在自然环境下生长的鸡可以自由漫步觅食，吃种子、虫子和餐厨垃圾。你的朋友更喜欢吃哪一种呢？

学科知识

Academic Knowledge

生物学	被驯化了的禽类通常被称为家禽；首个鸟类基因组的破译来自对鸡的研究；甲状腺肿大是由于饮食中缺少碘。
化　学	一种叫作氨丙啉的化学品可以防止球虫病，这是一种影响鸡的体内营养吸收的肠疾病；羊奶中钾、硒、钙的含量是牛奶的两倍；乳糖是牛奶中的糖分；碘是甲状腺功能所必需的微量无机物。
物　理	海藻比竹子生长更快，因为不需要克服地心引力；鸡蛋需要每天翻转3–5次以防止生长畸形。
工程学	杂交鸡的基因工程中，其中一个品种用于产肉（肥胖基因），另一个品种用于产蛋；埃及人掌握了人工孵化的技术，使母鸡有更多时间用于产蛋。
经济学	为了追求高效，人们会制造非常干的鸡饲料，这样鸡就得喝大量的水，这增加了在拥挤的鸡笼里通过水造成污染的风险，进而会因过量使用抗生素而影响人的免疫系统，付出额外的代价；两种养鸡产业的发展：一种是养殖纯种鸡，基本上与100年前相同（作为兴趣来养殖），一种是利用遗传学的商业化发展养鸡业；肉用和蛋用鸡饲料的高效生产；鸡粪作为动物性饲料和肥料的循环。
伦理学	现代化大规模生产使养鸡成为一个产业，而不考虑动物的权益。
历　史	鸡的驯养始于公元前2000年；鸡的起源可追溯至东南亚丛林；最初人们驯养鸡并不是为了鸡肉或鸡蛋，而是为了运动（斗鸡），这才是当时广泛养鸡的原因；在埃及的寺庙中悬挂鸡蛋是为了祈求充足的水源；在古波斯琐罗亚斯德教的信仰中，公鸡是吉利的象征，在黎明报晓啼叫，预示着宇宙中黑暗与光明交替的转折点；中国东北考古遗址发现鸡早在公元前5400年就出现了。
地　理	鸡起源于印度东北部至菲律宾群岛一带；布雷斯鸡是法国优质品种。
数　学	投入/产出表格：计算所有的投入，得出期待的产出值，包括除去不需要的副产品的花费。
生活方式	人们为了减少脂肪摄取由红肉改为吃白肉；动物权益的重要性日益突出；作为人们喜爱的甜点的冰淇淋文化；缺乏运动和锻炼会导致糖尿病和心脏衰竭。
社会学	关注效能与成本，忽视对人及动物的生活质量造成的影响；人们禁止斗鸡，认为这样对动物很残忍，可是大多数人并不认为生产肉食鸡是残忍的；母鸡被普遍作为养育及多产的象征。
心理学	人们抱怨困扰他们的事情却不去解决它；人们认为斗鸡表演中所表现出来的本能侵略性是非常鼓舞人心的，而不是毫无意义的、令人沮丧的；形容某人胆小如"鸡"；神经性焦虑。
系统论	需要调节食物生产，以适应地球上数十亿人的需求和传统供应链的生产力，而生产力不得不借转基因食物（如杂交鸡）来得到提高。

情感智慧

Emotional Intelligence

公　鸡

公鸡对农民给予他的食物以及对待他的方式感到不满，他抱怨农民唯一感兴趣的是他的肉产量，而不关心他的生活质量。他不同意把鸡按照产肉和产蛋加以区分，却对此无能为力。公鸡知识渊博，并且意识到如果不按照农民的意愿来工作，后果会非常不利。在设定生产目标时，他能够看到问题的细节。公鸡谈到人工孵化的小鸡永远见不到他们的母亲时变得感性。黄蜂催促他采取行动。公鸡一开始从自身需求出发，提出一个简单的要求。他知道哪些东西对他有好处，哪些对养鸡行业的产品质量有好处。他很关心自己的血脉传承。

黄　蜂

黄蜂不明白公鸡在抱怨什么。起初黄蜂不理解公鸡为什么抱怨，于是就跟公鸡开玩笑，并提出好笑的建议。当黄蜂明白母鸡所经受的苦难——必须每天都产一枚蛋时，才意识到事关生死。她对母鸡因无法完成工作量而受到的残酷惩罚感到惊讶，认为根本不可能达到这个标准。黄蜂催促公鸡停止抱怨，开始行动。起初黄蜂不理解海藻的重要性，但是后来她了解到海藻能提升产品（鸡蛋）质量，并在加入其他质量要素（羊奶）后改善其衍生产品（冰淇淋）的口感。

艺术

The Arts

你听说过艺术家安迪·沃霍尔吗？他的著名艺术作品之一就是汤罐头！让我们也来与朋友们一起创造一系列与鸡有关的不同颜色和背景的画作。首先简单勾勒出鸡的头部轮廓。可以在纸上也可以在电脑上画。接下来开始填充不同的颜色。诀窍就是不要使用单一颜色，而是用多种颜色的组合，这样就会有十多种对鸡的头部的不同诠释。比较一下你与你朋友的作品，看看谁使用了不同的颜色组合。就像自然界有生物多样性，你能看到艺术与文化的多样性吗？

思维拓展

Systems: Making the Connections

世界人口日益增长，需要不断生产更多的粮食来满足每个人的需求。50年前鸡肉还只被用于款待宾客，如今已经成为世界各地主要的肉类来源。斗鸡是已知的最古老的运动，鸡类也随之遍布各大洲。鸡的普及与3000年前埃及人发明人工孵化鸡蛋的技术有关。养鸡业随后在世界各地快速发展。然而，商业竞争追求高产量，与自动化及电子产业的工业化及标准化类似，只关注经济规模的扩大及成本的降低，忽略了对人类及动物造成的影响。

今天超市在售的99%鸡肉都是现代科学的产物，其产量惊人，超出了生物性能的极限。鸡肉的味道和肉质变得不如以前，脂肪含量也超出既定标准。鸡的肥胖症甚至已经对人类肥胖造成威胁。越来越多的人开始质疑这些养殖手段，以及对动物权益的忽视。这些禽类需要更多的尊重，更大的空间以及更少的压力。对于鸡饲料的质疑也开始出现，过去，鸡以昆虫、蟑螂、青草和种子为食，可如今它们的主食包括蛋白质和高脂谷类，这些在之前的饮食中都未出现过。这引发了关于我们以何种方式喂养世界的新讨论，我们以极低的成本，导致无法预料也无意造成的后果，然后去修复这些对禽类生活质量及人类健康所造成的破坏。当前主要的挑战在于降低成本的方法仅适用于主流产品，更高的时代价值和对品质的推崇，很快会被解读为更高的价格，它们本身并不是商业战略的一部分。

动手能力

Capacity to Implement

假设你有三四只出生几天的小鸡，把它们养在家中。它们小时候很容易养，但是一旦它们长成，你们所有人的生活可能会发生改变！你需要注意的第一件事就是区分公鸡和母鸡。如果你有一只公鸡，那么留意它早晨打鸣。一旦公鸡开始把每个人都叫醒，这很可能表示它将不得不住出去。你能养母鸡多久呢？你餐桌上的残渣是否足够它们吃？你有耐心等到它们第一回下蛋吗？一旦你有了这些经历，你就再也不会用与之前相同的方式对待鸡了。